# JOHN COUCH ADAMS

## *and the Discovery of Neptune*

J. C. ADAMS

# JOHN COUCH ADAMS

## *and the Discovery of Neptune*

BY

SIR HAROLD SPENCER JONES, F.R.S.
*The Astronomer Royal*

CAMBRIDGE
AT THE UNIVERSITY PRESS
1947

## CAMBRIDGE
### UNIVERSITY PRESS

University Printing House, Cambridge CB2 8BS, United Kingdom

Published in the United States of America by Cambridge University Press, New York

Cambridge University Press is part of the University of Cambridge.

It furthers the University's mission by disseminating knowledge in the pursuit of education, learning and research at the highest international levels of excellence.

www.cambridge.org
Information on this title: www.cambridge.org/9781107691896

First published 1947
Re-issued 2014

*A catalogue record for this publication is available from the British Library*

ISBN 978-1-107-69189-6 Paperback

Cambridge University Press has no responsibility for the persistence or accuracy of URLs for external or third-party internet websites referred to in this publication, and does not guarantee that any content on such websites is, or will remain, accurate or appropriate.

The portrait of J. C. Adams reproduced in the frontispiece was engraved by Samuel Cousins, A.R.A., after the painting by Thomas Mogford, 1851. Adams's Memorandum of October 1845 is reproduced in facsimile on pp. 15–17.

# JOHN COUCH ADAMS
## AND THE
## DISCOVERY OF NEPTUNE

O N the night of 13 March 1781 William Herschel, musician by profession but assiduous observer of the heavens in his leisure time, made a discovery that was to bring him fame. He had for some time been engaged upon a systematic and detailed survey of the whole heavens, using a 7 in. telescope of his own construction; he carefully noted everything that appeared in any way remarkable. On the night in question, in his own words:

'In examining the small stars in the neighbourhood of H Geminorum I perceived one that appeared visibly larger than the rest; being struck with its uncommon appearance I compared it to H Geminorum and the small star in the quartile between Auriga and Gemini, and finding it so much larger than either of them, I suspected it to be a comet.'

Most observers would have passed the object by without noticing anything unusual about it, for the minute disk was only about 4 sec. in diameter. The discovery was made possible by the excellent quality of Herschel's telescope, and by the great care with which his observations were made.

The discovery proved to be of greater importance than Herschel suspected, for the object he had found was not a comet, but a new planet, which revolved round the Sun in a nearly circular path at a mean distance almost exactly double that of Saturn; it was unique, because no planet had ever before been discovered; the known planets, easily visible to the naked eye, did not need to be discovered.

After the discovery of Uranus, as the new planet was called, it was ascertained that it had been observed as a star and its position recorded on a score of previous occasions. The earliest of these observations was made by Flamsteed at Greenwich in 1690. Lemonnier in 1769 had observed its transit six times in the course of 9 days and, had he compared the observations with one another, he could not have failed to anticipate Herschel in the discovery. As Uranus takes 84 years to make a complete revolution round the Sun, these earlier observations were of special value for the investigation of its orbit.

The positions of the planet computed from tables constructed by Delambre soon began to show discordances with observation, which became greater as time went on. As there might have been error or incompleteness in Delambre's theory and tables, the task of revision was undertaken by Bouvard, whose tables of the planet appeared in 1821. Bouvard found that, when every correction for the perturbations in the motion of Uranus by the other planets was taken into account, it was not possible to reconcile the old observations of Flamsteed, Lemonnier, Bradley, and Mayer with the observations made subsequently to the discovery of the planet in 1781.

'The construction of the tables, then,' said Bouvard, 'involves this alternative: if we combine the ancient observations with the modern, the former will be sufficiently well represented, but the latter will not be so, with all the precision which their superior accuracy demands; on the other hand, if we reject the ancient observations altogether, and retain only the modern, the resulting tables will faithfully conform to the modern observations, but will very inadequately represent the more ancient. As it was necessary to decide between these two courses, I have adopted the latter, on the ground that it unites the greatest number of probabilities in favour of the truth, and I leave to the future the task of discovering whether the difficulty of reconciling the two systems is connected with the ancient observations, or whether it depends on some foreign and unperceived cause which may have been acting upon the planet.'

Further observations of Uranus were for a time found to be pretty well represented by Bouvard's Tables, but systematic discordances between observations and the tables gradually began to show up. As time went on, observations continued to deviate more and more from the tables. It began to be suspected that there might exist an unknown distant planet, whose gravitational attraction was disturbing the motion of Uranus. An alternative suggestion was that the inverse square law of gravitation might not be exact at distances as great as the distance of Uranus from the Sun.

The problem of computing the perturbations in the motion of one planet by another moving planet, when the undisturbed orbits and the masses of the planets are known is fairly straightforward, though of some mathematical complexity. The inverse problem, of analysing the perturbations in the motion of one planet in order to deduce the position, path and mass of the planet which is producing these perturbations, is of much greater complexity and difficulty. A little consideration will, I think, show that this must be so. If a planet were exposed solely to the attractive influence of the Sun, its orbit would be an ellipse. The attractions of the other planets perturb its motion and cause it to deviate now on the one side and now on the other side of this ellipse. To determine the elements of the elliptic orbit from the positions of the planet as assigned by observation, it is necessary first to compute the perturbations produced by the other planets and to subtract them from the observed positions.

The position of the planet in this orbit at any time, arising from its undisturbed motion, can be calculated; if the perturbations of the other planets are then computed and added, the true position of the planet is obtained. The whole procedure is, in practice, reduced to a set of tables. But if Uranus is perturbed by a distant *unknown* planet, the observed positions when corrected by the subtraction of the perturbations caused by the *known* planets are not the positions in the true elliptic orbit; the perturbations of the unknown planet have not been allowed for. Hence when the corrected positions are analysed in order to determine the elements of the elliptic orbit, the derived elements will be falsified. The positions of Uranus computed from tables such as Bouvard's would be in error for two reasons; in the first place, because they are based upon incorrect elements of the elliptic orbit; in the second place, because the perturbations produced by the unknown planet have not been applied. The two causes of error have a common origin and are inextricably entangled in each other, so that neither can be investigated independently of the other. Thus though many astronomers thought it probable that Uranus was perturbed by an undiscovered planet, they could not prove it. No occasion had arisen for the solution of the extremely complicated problem of what is termed inverse perturbations, starting with the perturbed positions and deducing from them the position and motion of the perturbing body.

The first solution of this intricate problem was made by a young Cambridge mathematician, John Couch Adams. As a boy at school Adams had shown conspicuous mathematical ability, an interest in astronomy, and skill and accuracy in numerical computation. At the age of 16 he had computed the circumstances of an annular eclipse of the Sun, as visible from Lidcot, near Launceston, where his brother lived. He entered St John's College in October, 1839, at the age of 20, and in 1843 graduated Senior Wrangler, being reputed to have obtained more than double the marks awarded to the Second Wrangler. In the same year he became first Smith's Prizeman and was elected Fellow of his College.

Whilst still an undergraduate his attention had been drawn to the irregularities in the motion of Uranus. After his death there was found among his papers this memorandum, written at the beginning of his second long vacation:

> *Memoranda.*
>
> *1841, July 3. Formed a design, in the beginning of this week, of investigating, as soon as possible after taking my degree, the irregularities in the motion of Uranus, which are yet unaccounted for; in order to find whether they may be attributed to the action of an undiscovered planet beyond it; and if possible hence to determine the elements of its orbit, &c. approximately, which would probably lead to its discovery.*

'1841, July 3. Formed a design in the beginning of this week, of investigating, as soon as possible after taking my degree, the irregularities in the motion of Uranus, which are yet unaccounted for; in order to find whether they may be attributed to the action of an undiscovered planet beyond it; and if possible thence to determine the elements of its orbit, etc. approximately, which would probably lead to its discovery.'

As soon as Adams had taken his degree he attempted a first rough solution of the problem, with the simplifying assumptions that the unknown planet moved in a circular orbit, in the plane of the orbit of Uranus, and that its distance from the Sun was twice the mean distance of Uranus, this being the distance to be expected according to the empirical law of Bode. This preliminary solution gave a sufficient improvement in the agreement between the corrected theory of Uranus and observation to encourage him to pursue the investigation further. In order to make the observational data more complete application was made in February 1844 by Challis, the Plumian Professor of Astronomy, to Airy, the Astronomer Royal, for the errors of longitude of Uranus for the years 1818–26. Challis explained that he required them for a young friend, Mr Adams of St John's College, who was working at the theory of Uranus. By return of post, Airy sent the Greenwich data not merely for the years 1818–26 but for the years 1754–1830.

Adams now undertook a new solution of the problem, still with the assumption that the mean distance of the unknown planet was twice that of Uranus but without assuming the orbit to be circular. During term-time he had little opportunity to pursue his investigations and most of the work was undertaken in the vacations. By September 1845, he had completed the solution of the problem, and gave to Challis a paper with the elements of the orbit of the planet, as well as its mass and its position for 1 October 1845. The position indicated by Adams was actually within 2° of the position of Neptune at that time. A careful search in the vicinity of this position should have led to the discovery of Neptune. The comparison between observation and theory was satisfactory and Adams, confident in the validity of the law of gravitation and in his own mathematics, referred to the 'new planet'.

Challis gave Adams a letter of introduction to Airy, in which he said that 'from his character as a mathematician, and his practice in calculation, I should consider the deductions from his premises to be made in a trustworthy manner'. But the Astronomer Royal was in France when Adams called at Greenwich. Airy, immediately on his return, wrote to Challis saying: 'would you mention to Mr Adams that I am very much interested with the subject of his investigations, and that I should be delighted to hear of them by letter from him?'

Towards the end of October Adams called at Greenwich, on his way from Devonshire to Cambridge, on the chance of seeing the Astronomer Royal. At about that time Airy was occupied almost every day with meetings of the Railway Gauge Commission and he was in London when Adams called. Adams left his card and said that he would call again. The card was taken to Mrs Airy, but she was not told of the intention of Adams to call later. When Adams made his second call, he was informed that the Astronomer Royal was at dinner; there was no message for him and he went away feeling mortified. Airy, unfortunately, did not know of this second visit at the time. Adams left a paper summarizing the results which he had obtained and giving a list of the residual errors of the mean longitude of Uranus, after taking account of the disturbing action of the new planet. These errors were satisfactorily small, except for the first observation by Flamsteed in 1690. A few days later Airy wrote to Adams acknowledging the paper and enquiring whether the perturbations would explain

14

1845 October

According to my calculations, the ob? irregularities in the motion of Uranus may be accounted for, by supposing the existence of an ext<sup>r</sup> planet; the mass & orbit of wh. are as follows

Mean Dist. (assumed nearly in accordance with Bode's law)

38.4

Mean sid<sup>l</sup> motn in 365.25 days

1°. 30.9

Mean Long. 1<sup>st</sup> Oct<sup>r</sup> 1845

323°.34′

Long. Perih<sup>n</sup>

315.55

Eccent?

0.1610

Mass (that of Sun being unity)

0.00016.56

3

For the modern obs.^ns I have
used the method of Normal places,
taking the mean of the Tabular
errors as given by obs.^ns near 3
consecutive opp.^ns to correspond
with the mean of the times & the
Greenwich obs.^ns have been used down
to 1830, since wh. the Cambridge &
Greenwich obs.^ns and those given
in the Ashton. Nachr. have been
made use of. The foll.^g are the
sum? errors of mean Longitude

| obs.—Theory | | obs.—Theory | | obs.—Theory | |
|---|---|---|---|---|---|
| 1780 | +0".27 | 1801 | −0".04 | 1822 | +0".30 |
| 1783 | −0.23 | 1804 | +1.76 | 1825 | +1.92 |
| 1786 | −0.96 | 1807 | −0.21 | 1828 | +2.25 |
| 1789 | +1.82 | 1810 | +0.56 | 1831 | −1.06 |
| 1792 | −0.91 | 1813 | −0.94 | 1834 | −1.44 |
| 1795 | +0.09 | 1816 | −0.31 | 1837 | −1.62 |
| 1798 | −0.99 | 1819 | −2.00 | 1840 | +1.73 |

The error for 1780 is concluded from
that for 1781 given by obs.^n compared
with those of 4 or 5 following years
& also with Lemonnier's obs.^ns in 1769
& 1771.

For the ancient obs.ⁿˢ the foll⁸. are the remᵍ. errors

| Obsⁿ. – Theory | | Obsⁿ. – Theory | |
|---|---|---|---|
| 1690 | + 44.″4 | 1756 | – 4.″0 |
| 1712 | + 6.7 | 1763 | – 5.1 |
| 1715 | – 6.8 | 1769 | + 0.6 |
| 1750 | – 1.6 | 1771 | + 11.8 |
| 1753 | + 5.7 | | |

The errors are small except for Flamsteed's obsⁿ of 1690. This being an isolated obsⁿ very distant from the rest, I thought it best not to use it in forming the —ⁿˢ of condⁿ. It is not improbable however that this error might be destʳᵈ by a small change in the assumed mean motion of the new planet.

                    J. C. Adams

the errors of the radius vector of Uranus as well as the errors of longitude; in the reduction of the Greenwich observations, Airy had shown that not only the longitude of Uranus but also its distance from the Sun (called the radius vector) showed discordances from the tabular values. Airy said at a later date that he waited with much anxiety for the answer to this query, which he looked upon as an *experimentum crucis*, and that if Adams had replied in the affirmative, he would at once have exerted all his influence to procure the publication of Adams's theory. It should be emphasized that neither Challis nor Airy knew anything about the details of Adams's investigation. Adams had attacked this difficult problem entirely unaided and without guidance. Confident in his own mathematical ability he sought no help and he needed no help.

Adams never replied to the Astronomer Royal's query; but for this failure to reply, he would almost certainly have had the sole glory of the discovery of Neptune. Airy and Adams were looking at the same problem from different points of view; Adams was so convinced that the discordances between the theory of Uranus and observation were due to the perturbing action of an unknown planet that no alternative hypothesis was considered by him; Airy,

18

on the other hand, did not exclude the possibility that the law of gravitation might not apply exactly at great distances. The purpose of his query, to which he attached great importance, was to decide between the two possibilities. As he later wrote to Challis (21 December 1846):

'There were two things to be explained, which might have existed each independently of the other, and of which one could be ascertained independently of the other: viz. error of longitude and error of radius vector. And there is no *a priori* reason for thinking that a hypothesis which will explain the error of longitude will also explain the error of radius vector. If, after Adams had satisfactorily explained the error of longitude he had (with the numerical values of the elements of the two planets so found) converted his formula for perturbation of radius vector into numbers, and if these numbers had been discordant with the *observed* numbers of discordances of radius vector, then *the theory would have been false,* NOT from any error of Adams's BUT from a failure in the law of gravitation. On this question therefore turned the continuance or fall of the law of gravitation.'

What were the reasons for Adams's failure to reply? There were several; he gave them himself at a later date (18 November 1846) in a letter to Airy. He wrote as follows:

'I need scarcely say how deeply I regret the neglect of which I was guilty in delaying to reply to the question respecting the Radius Vector of Uranus, in your note of November 5th, 1845. In palliation, though not in excuse of this neglect, I may say that I was not aware of the importance which you attached to my answer on this point and I had not the smallest notion that you felt any difficulty on it. ...For several years past, the observed place of Uranus has been falling more and more rapidly behind its tabular place. In other words, the real angular motion of Uranus is considerably *slower* than that given by the Tables. This appeared to me to show clearly that the Tabular Radius Vector would be considerably increased by any Theory which represented the motion in Longitude.... Accordingly, I found that if I simply corrected the elliptic elements, so as to satisfy the modern observations as nearly as possible without taking into account any additional perturbations, the corresponding increase in the Radius Vector would not be very different from that given by my actual Theory. Hence it was that I waited to defer writing to you till I could find time to draw up an account of the method employed to obtain the results which I had communicated to you. More than once I commenced writing with this object, but unfortunately did not persevere. I was also much pained at not having been able to see you when I called at the Royal Observatory the second time, as I felt that the whole matter might be better explained by half an hour's conversation than by several letters, in writing which I have always experienced a strange difficulty. I entertained from the first the strongest conviction that the observed anomalies were due to the action of an exterior planet; no other hypothesis appeared to me to possess the slightest claims to attention. Of the accuracy of my calculations I was quite sure, from the care with which they were made and the number of times I had examined them. The only point which appeared to admit of any doubt was the assumption as to the mean distance and this I soon proceeded to correct. The work however went on very slowly throughout, as I had scarcely any time to give to these investigations, except during the vacations.

'I could not expect, however, that practical astronomers, who were already fully occupied with important labours, would feel as much confidence in the results of my investigation, as I myself did;

and I therefore had our instruments put in order, with the express purpose, if no one else took up the subject, of undertaking the search for the planet myself, with the small means afforded by our observatory at St John's.'

Airy was a man with a precise and orderly mind, extremely methodical and prompt in answering letters. Another person might have followed the matter up, but not Airy. In a letter of later date to Challis, he said that 'Adams's silence . . . was so far unfortunate that it interposed an effectual barrier to all further communication. It was clearly impossible for me to write to him again.'

Meanwhile, another astronomer had turned his attention to the problem of accounting for the anomalies in the motion of Uranus. In the summer of 1845 Arago, Director of the Paris Observatory, drew the attention of his friend and protégé, Le Verrier, to the importance of investigating the theory of Uranus. Le Verrier was a young man, 8 years older than Adams, with an established reputation in the astronomical world, gained by a brilliant series of investigations in celestial mechanics. In contrast, Adams was unknown outside the circle of his Cambridge friends and he had not yet published anything.

Le Verrier decided to devote himself to the problem of Uranus and laid aside some researches on comets, on which he had been engaged. His investigations received full publicity, for the results were published, as the work proceeded, in a series of papers in the *Comptes Rendus* of the French Academy. In the first of these, communicated in November 1845 (a month after Adams had left his solution of the problem with the Astronomer Royal), Le Verrier recomputed the perturbations of Uranus by Jupiter and Saturn, derived new orbital elements for Uranus, and showed that these perturbations were not capable of explaining the observed irregularities of Uranus. In the next paper, presented in June 1846, Le Verrier discussed possible explanations of the irregularities and concluded that none was admissible, except that of a disturbing planet exterior to Uranus. Assuming, as Adams had done, that its distance was twice the distance of Uranus and that its orbit was in the plane of the ecliptic, he assigned its true longitude for the beginning of 1847; he did not obtain the elements of its orbit nor determine its mass.

The position assigned by Le Verrier differed by only 1° from the position which Adams had given seven months previously. Airy now felt no doubt about the accuracy of both calculations; he still required to be satisfied about the error of the radius vector, however, and he accordingly addressed to Le Verrier the query that he had addressed to Adams, but this time in a more explicit form. He asked whether the errors of the tabular radius vector were the consequence of the disturbance produced by an exterior planet, and explained why, by analogy with the moon's variation, this did not seem to him necessarily to be so. Le Verrier replied a few days later giving an explanation which Airy found completely satisfactory. The errors of the tabular radius vector, said Le Verrier, were not produced actually by the disturbing planet; Bouvard's orbit required correction, because it had been based on positions which were not true elliptic positions, including, as they did, the perturbations by the outer planet; the correction of the orbit, which was needed on this account, removed the discordance between the observed and tabular radius vector.

Airy was a man of quick and incisive action. He was now fully convinced that the true explanation of the irregularities in the motion of Uranus had been provided and he felt confident that the new planet would soon be found. He had already, a few days before receiving the reply from Le Verrier,

informed the Board of Visitors of the Royal Observatory, at their meeting in June, of the extreme
probability of discovering a new planet in a very
short time. It was in consequence of this strongly
expressed opinion of Airy that Sir John Herschel
(a member of the Board) in his address on 10 September to the British Association, at its meeting at
Southampton, said: 'We see it [a probable new
planet] as Columbus saw America from the shores
of Spain. Its movements have been felt, trembling
along the far-reaching line of our analysis, with a
certainty hardly inferior to that of ocular demonstration.'

Airy considered that the most suitable telescope
with which to make the search for the new planet
was the Northumberland telescope of the Cambridge Observatory, which was larger than any
telescope at Greenwich and more likely to detect a
planet whose light might be feeble. Airy offered to
lend Challis one of his assistants, if Challis was too
busy to undertake the search himself. He pointed
out that the most favourable time for the search
(when the undiscovered planet would be at opposition) was near at hand. A few days later, Airy sent
Challis detailed directions for carrying out the
search and in a covering letter said that, in his

opinion, the importance of the inquiry exceeded that of any current work, which was of such a nature as not to be totally lost by delay.

Challis decided to prosecute the search himself and began observing on 29 July 1846, three weeks before opposition. The method adopted was to make three sweeps over the area to be searched, mapping the positions of all the stars observed, and completing each sweep before beginning the next. If the planet was observed it would be revealed, when the different sweeps were compared, by its motion relative to the stars.

What followed was not very creditable to Challis. He started by observing in the region indicated by Adams: the first four nights on which observations were made were 29 July, 30 July, 4 August and 12 August. But no comparison was made, as the search proceeded, between the observations on different nights. He did indeed make a partial comparison between the nights of 30 July and 12 August, merely to assure himself that the method of observation was adequate. He stopped short at No. 39 of the stars observed on 12 August; as he found that all these had been observed on 30 July, he felt satisfied about the method of observation. If he had continued the comparison for another ten stars he would have found that a star of the 8th magnitude observed on 12 August was missing in the series of 30 July. This was the planet: it had wandered into the zone between the two dates. Its discovery was thus easily within his grasp. But 12 August was not the first time on which Challis had observed the planet; he had already observed it on 4 August and if he had compared the observations of 4 August with the observations of either 30 July or of 12 August, the planet would have been detected.

When we recall Airy's strong emphasis on carrying on the search in preference to any current work, Challis's subsequent excuses to justify his failure were pitiable. He had delayed the comparisons, he said, partly from being occupied with comet reductions (which could well have waited), and partly from a fixed impression that a long search was required to ensure success. He confessed that, in the whole of the undertaking, he had too little confidence in the indications of theory. Oh! man of little faith! If only he had shared Airy's conviction of the great importance of the search.

But we have anticipated somewhat. While Challis was laboriously continuing his search, Adams wrote on 2 September an important letter to Airy who, unknown to Adams, was then in Germany. He referred to the assumption in his first calculations that the mean distance of the supposed disturbing planet was twice that of Uranus. The investigation, he said, could scarcely be considered satisfactory while based on anything arbitrary. He had therefore repeated his calculations, assuming a somewhat smaller mean distance. The result was very satisfactory in that the agreement between theory and observations was somewhat improved and, at the same time, the eccentricity of the orbit, which in the first solution had an improbably large value, was reduced. He gave the residuals for the two solutions, and remarked that the comparison with recent Greenwich observations suggested that a still better agreement could be obtained by a further reduction in the mean distance. He asked for the results of the Greenwich observations for 1844 and 1845. He then gave the corrections to the tabular radius vector of Uranus and remarked that they were in close agreement with those required by the Greenwich observations.

Two days earlier, on 31 August, Le Verrier had communicated a third paper to the French Academy which was published in a number of the *Comptes Rendus* that reached England near the end of September. Challis received it on 29 September.

Le Verrier gave the orbital elements of the hypothetical planet, its mass, and its position. From the mass and distance of the planet he inferred, on the reasonable assumption that its mean density was equal to the mean density of Uranus, that it should show a disk with an angular diameter of about 3·3 sec. Le Verrier went on to remark as follows:

'It should be possible to see the new planet in good telescopes and also to distinguish it by the size of its disk. This is a very important point. For if the planet could not be distinguished by its appearance from the stars it would be necessary, in order to discover it, to examine all the small stars in the region of the sky to be explored, and to detect a proper motion of one of them. This work would be long and wearisome. But if, on the contrary, the planet has a sensible disk which prevents it from being confused with a star, if a simple study of its physical appearance can replace the rigorous determination of the positions of all the stars, the search will proceed much more rapidly.'

After reading this memoir on 29 September, Challis searched the same night in the region indicated by Le Verrier (which was almost identical with that indicated by Adams, in the first instance, a year earlier), looking out particularly for a visible disk. Of 300 stars observed he noted one and one only as seeming to have a disk. This was, in actual fact, the planet. Its motion might have been detected in the course of a few hours, but Challis waited for confirmation until the next night, when no observation was possible because the Moon was in the way. On 1 October he learnt that the planet had been discovered at Berlin on 23 September. His last chance of making an independent discovery had gone.

For on 23 September Galle, Astronomer at the Berlin Observatory, had received a letter from Le Verrier suggesting that he should search for the unknown planet, which would probably be easily distinguished by a disk. D'Arrest, a keen young volunteer at the Observatory, asked to share in the search, and suggested to Galle that it might be worth looking among the star charts of the Berlin Academy, which were then in course of publication, to verify whether the chart for Hour 21 was amongst those that were finished. It was found that this chart had been printed at the beginning of 1846, but had not yet been distributed; it was therefore available only to the astronomers at the Berlin Observatory. Galle took his place at the telescope, describing the configurations of the stars he saw, while d'Arrest followed them on the map, until

Galle said: 'And then there is a star of the 8th magnitude in such and such a position', whereupon d'Arrest exclaimed: 'That star is not on the map.' An observation the following night showed that the object had changed its position and proved that it was the planet. Had this chart been available to Challis, as it would have been but for the delay in distribution, he would undoubtedly have found the planet at the beginning of August, some weeks before Le Verrier's third memoir was presented to the French Academy.

On 1 October, Le Verrier wrote to Airy informing him of the discovery of the planet. He mentioned that the Bureau des Longitudes had adopted the name Neptune, the figure a trident, and that the name Janus (which had also been suggested) would have the inconvenience of making it appear that the planet was the last in the solar system, which there was no reason to believe.

The discovery of the planet, following the brilliant researches of Le Verrier, which were known to the scientific world through their publication by the French Academy, was received with admiration and delight, and was acclaimed as one of the greatest triumphs of the human intellect. The prior investigations of Adams, his prediction of the position of the

5

planet, the long patient search by Challis were known to only a few people in England. Adams had published nothing; he had communicated his results to Challis and to Airy, but neither of them knew anything of the details of his investigations; his name was unknown in astronomical circles outside his own country. Adams had actually drawn up a paper to be read at the meeting of the British Association at Southampton early in September, but he did not arrive in sufficient time to present it, as Section A closed its meetings one day earlier than he had expected.

The first reference in print to the fact that Adams had independently reached conclusions similar to those of Le Verrier was made in a letter from Sir John Herschel, published in the *Athenaeum* of 3 October. It came as a complete surprise to the French astronomers and ungenerous aspersions were cast upon the work of Adams. It was assumed that his solution was a crude essay which would not stand the test of rigorous examination and that, as he had not published any account of his researches, he could not establish a claim to priority or even to a share in the discovery. Some justification seemed to be afforded by an unfortunate letter from Challis to Arago, of 5 October, stating that he had searched for the planet, in conformity with the suggestions of Le Verrier, and had observed an object on 29 September which appeared to have a disk and which later was proved to have been the planet. No reference was made in this letter to the investigations of Adams or to his own earlier searches during which the planet had twice been observed. Airy, moreover, wrote to Le Verrier on 14 October, mentioning that collateral researches, which had led to the same result as his own, had been made in England. He went on to say: 'If in this I give praise to others I beg that you will not consider it as at all interfering with my acknowledgment of your claims. You are to be recognized, without doubt, as the real predictor of the planet's place. I may add that the English investigations, as I believe, were not so extensive as yours. They were known to me earlier than yours.' It is difficult to understand why Airy wrote in these terms; he had expressed the highest admiration for the manner in which the problem had been solved by Le Verrier, but he was not in a position to express any opinion about the work of Adams, which he had not yet seen.

At the meeting of the French Academy on 12 October, Arago made a long and impassioned defence of his protégé, Le Verrier, and a violent attack on Adams, referring scornfully to what he described as his *clandestine* work. 'Le Verrier is to-day asked to share the glory, so loyally, so rightly earned, with a young man who has communicated nothing to the public and whose calculations, more or less incomplete, are, with two exceptions, totally unknown in the observatories of Europe! No! no! the friends of science will not allow such a crying injustice to be perpetrated.' He concluded by saying that Adams had no claim to be mentioned, in the history of Le Verrier's planet, by a detailed citation nor even by the slightest allusion. National feeling ran very high in France. The paper *Le National* asserted that the three foremost British astronomers (Herschel, Airy and Challis) had organized a miserable plot to steal the discovery from M. Le Verrier and that the researches of Adams were merely a myth invented for this purpose.

In England opinion was divided; some English astronomers contended that because Adams's results had not been publicly announced he could claim no share in the discovery. But for the most part it was considered that the credit for the successful prediction of the position of the unknown planet should be shared equally between Adams and Le Verrier. Adams himself took no part in the heated discussions which went on for some time with regard to the credit for the discovery of the new planet; he never uttered a single word of criticism or blame in connexion with the matter.

The controversy was lifted to a higher plane by a letter from Sir John Herschel to *The Guardian* in which he said:

'The history of this grand discovery is that of *thought* in one of its highest manifestations, of science in one of its most refined applications. So viewed, it offers a deeper interest than any personal question. In proportion to the importance of the step, it is surely interesting to know that more than one mathematician has been found capable of taking it. The fact, thus stated, becomes, so to speak, a measure of the maturity of our science; nor can I conceive anything better calculated to impress the general mind with a respect for the mass of accumulated facts, laws, and methods, as they exist at present, and the reality and efficiency of the forms into which they have been moulded, than such a circumstance. We need some reminder of this kind in England, where a want of faith in the higher theories is still to a certain degree our besetting weakness.'

At the meeting of the Royal Astronomical Society on 13 November 1846, three important papers were read. The first, by the Astronomer Royal, was an 'Account of some Circumstances historically connected with the Discovery of the Planet Exterior to Uranus'. All the correspondence with Adams, Challis and Le Verrier was given, as well as the two memoranda from Adams, the whole being linked together by Airy's own comments. The account made it perfectly clear that Adams and Le Verrier had independently solved the same problem, that the positions which they had assigned to the new planet were in close agreement, and that Adams had been the first to solve the problem. The second was Challis's 'Account of Observations undertaken in search of the Planet discovered at Berlin on September 23, 1846', which showed that in the course of the search for the planet, he had twice observed it before its discovery at Berlin, and that he had observed it a third time before the news of this discovery reached England. The third paper was by Adams and was entitled 'An Explanation of the observed Irregularities in the Motion of Uranus, on the Hypothesis of Disturbances caused by a more Distant Planet; with a determination of the Mass, Orbit, and Position of the Disturbing Body'.

Adams's memoir was a masterpiece; it showed a thorough grasp of the problem; a mathematical maturity which was remarkable in one so young; and a facility in dealing with complex numerical computations. Lieut. Stratford, Superintendent of the Nautical Almanac, reprinted it as an Appendix to the *Nautical Almanac* for 1851, then in course of publication, and sent sufficient copies to Schumacher, editor of the *Astronomische Nachrichten*, for distribution with that periodical. Hansen, the foremost exponent of the lunar theory, wrote to Airy to say that, in his opinion, Adams's investigation showed more mathematical genius than Le Verrier's. Airy, a competent judge, gave his own opinion in a letter to Biot, who had sent to Airy a paper he had written about the new planet. He sent it, he said, with some diffidence because he had expressed a more favourable opinion of the work of Adams than Airy had given. In reply, Airy wrote: 'On the whole I think his [Adams's] mathematical investigation superior to M. Le Verrier's. However, both are so admirable that it is difficult to say.' He went on to state that 'I believe I have done more than any other person to place Adams in his proper position'.

With the independent investigations of both men published, there was no difficulty in agreeing that each was entitled to an equal share of the honour. The verdict of history agrees with that of Sir John

Herschel who, in addressing the Royal Astronomical Society in 1848, said:

'As genius and destiny have joined the names of Le Verrier and Adams, I shall by no means put them asunder; nor will they ever be pronounced apart so long as language shall celebrate the triumphs of science in her sublimest walks. On the great discovery of Neptune, which may be said to have surpassed, by intelligible and legitimate means, the wildest pretensions of clairvoyance, it would now be quite superfluous for me to dilate. That glorious event and the steps which led to it, and the various lights in which it has been placed, are already familiar to everyone having the least tincture of science. I will only add that as there is not, nor henceforth ever can be, the slightest rivalry on the subject of these two illustrious men—as they have met as brothers, and as such will, I trust, ever regard each other—we have made, we could make, no distinction between them on this occasion. May they both long adorn and augment our science, and add to their own fame, already so high and pure, by fresh achievements.'

Although on 1 October, Le Verrier had informed Airy that the Bureau des Longitudes had assigned the name Neptune to the new planet, Arago announced to the French Academy on 5 October that Le Verrier had delegated to him the right of naming the planet and that he had decided, in the exercise of this right, to call it Le Verrier. 'I pledge myself', he said, 'never to call the new planet by any other name than Le Verrier's Planet.' As though to justify this name, Le Verrier's collected memoirs on the perturbations of Uranus, which were published in the *Connaissance des Temps* for 1849 were given the title 'Recherches sur les mouvements de la planete Herschel (dite Uranus)' with a footnote to say that Le Verrier considered it as a strict duty, in future publications, to ignore the name of Uranus entirely and to call the planet only by the name of Herschel!

The name Le Verrier for the planet was not welcomed outside France. It was not in accordance with the custom of naming planets after mythological deities, and it ignored entirely the claims of Adams. Moreover, it might set a precedent. As Smyth said to Airy: 'Mythology is neutral ground. Herschel is a good name enough. Le Verrier somehow or other suggests a Fabriquant and is therefore not so good. But just think how awkward it would be if the next planet should be discovered by a German, by a Bugge, a Funk, or your hirsute friend Boguslawski!'

The widespread feeling against the name of Le Verrier was shared in Germany by Encke, Gauss and Schumacher and in Russia by Struve. Airy therefore wrote to Le Verrier:

'From my conversation with lovers of astronomy in England and from my correspondence with astronomers in Germany, I find that the name assigned by M. Arago is not well received. They think, in the first place, that the character of the name is at variance with that of the names of all the other planets. They think in the next place that M. Arago, as your delegate, could do only what you could do, and that you would not have given the name which M. Arago has given. They are all desirous of receiving a mythological name selected by you. In these feelings I do myself share. It was believed at first that you approved of the name Neptune, and in that supposition we have used the name Neptune when it was necessary to give a name. Now if it was understood that you still approve of the name Neptune (or Oceanus as some English mythologists suggested—or any other of the same class), I am sure that all England and Germany would adopt it at once. I am not sure that they will adopt the name which M. Arago has given.'

Airy might have added, but did not, that there

was a general feeling, not merely in England, but also in Germany and Russia, that the name Le Verrier by implication denied any credit to the work of Adams and that, for this reason also, it was inappropriate.

Le Verrier, in reply, said that since one spoke of Comet Halley, Comet Encke, etc., he saw nothing inappropriate in Planet Le Verrier; that the Bureau des Longitudes had given the name Neptune without his consent and had now withdrawn it;* and that, since he had delegated the selection of the name to Arago, it was a matter that no longer concerned him. At a later date, when relations between Arago and Le Verrier had become strained, the true story was

* This statement by Le Verrier was not correct. The minutes of the Bureau des Longitudes show that the Bureau had not considered assigning a name by 1 October, when Le Verrier had written not only to Airy but also to various other astronomers in Germany and Russia informing them that the Bureau des Longitudes 'had adopted the name Neptune, the figure a trident'. The Bureau neither assigned the name Neptune nor subsequently withdrew it. The minutes of the Bureau des Longitudes show that Le Verrier's statements were repudiated by the Bureau at a subsequent meeting. It seems that the name Neptune was Le Verrier's own choice in the first instance but that he soon decided that he would like the planet to be named Le Verrier. There is no explanation of his reasons for stating that the name Neptune had been assigned by the Bureau des Longitudes; it was, in fact, outside the competence of the Bureau to assign a name to a newly discovered planet.

told by Arago. It appears that Arago had at first agreed to the name Neptune, but Le Verrier had implored him, in order to serve him as a friend and as a countryman, to adopt the name Le Verrier. Arago had in the end agreed, but on condition that Uranus should always be called Planet Herschel, a name which Arago himself had frequently used. The greatest men are liable to human weaknesses and failings; Le Verrier was described by his friends as a *mauvais coucheur*, an uncomfortable bedfellow. By the general concensus of astronomers the name Neptune was adopted for the new planet; the name Le Verrier did not long survive.

In the history of the discovery of Neptune so many chances were missed which might have changed completely the course of events, that it is perhaps not surprising to find that the planet might have been discovered 50 years earlier. When sufficient observations of Neptune had been obtained to enable a fairly accurate orbit to be computed, a search was made to find out whether the planet had been observed as a star before its discovery. It was discovered that a star recorded in the *Histoire Céleste* of Lalande as having been observed on 10 May 1795 was missing in the sky; its position was marked as uncertain but was in close agreement with the position to be expected for Neptune. The original manuscripts of Lalande at the Paris Observatory were consulted; it was found that Lalande had, in fact, observed Neptune not only on 10 May but also on 8 May. The two positions, being found discordant and thought to refer to one and the same star, Lalande rejected the observation of 8 May and printed in the *Histoire Céleste* only the observation of 10 May marking it as doubtful, although it was not so marked in the manuscript. The change in position in the two days agreed closely with the motion of Neptune in the interval. If Lalande had taken the trouble to make a further observation to check the other two, he could scarcely have failed to discover the planet. Airy's comment, when sending the information about the two observations to Adams, was 'Let no one after this blame Challis'.

Milton Keynes UK
Ingram Content Group UK Ltd.
UKHW032321161024
449665UK00001B/1